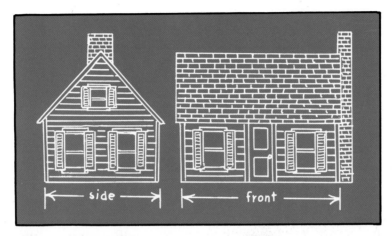

Building a House

by Byron Barton

Greenwillow Books, New York

Building a House. Copyright © 1981 by Byron Barton.
All rights reserved. Manufactured in China by South China
Printing Company Ltd. www.harperchildrens.com
First Edition 15 14
Library of Congress Cataloging-in-Publication Data:
Barton, Byron. Building a house. "Greenwillow Books."

se.
use
construction.] I. Title. TH4811.5.B37690'.8373 80-22674
ISBN 0-688-80291-5   ISBN 0-688-84291-7 (lib. bdg.)
ISB

On a green hill

a machine digs a big hole.

Builders hammer and saw.

A cement mixer pours cement.

Bricklayers lay large white blocks.

Carpenters come and make a wooden floor.

They put up walls.

They build a roof.

A bricklayer builds a fireplace and a chimney too.

A plumber puts in pipes for water.

An electrician wires for electric lights.

Painters paint inside and out.

The workers leave.

The house is built.

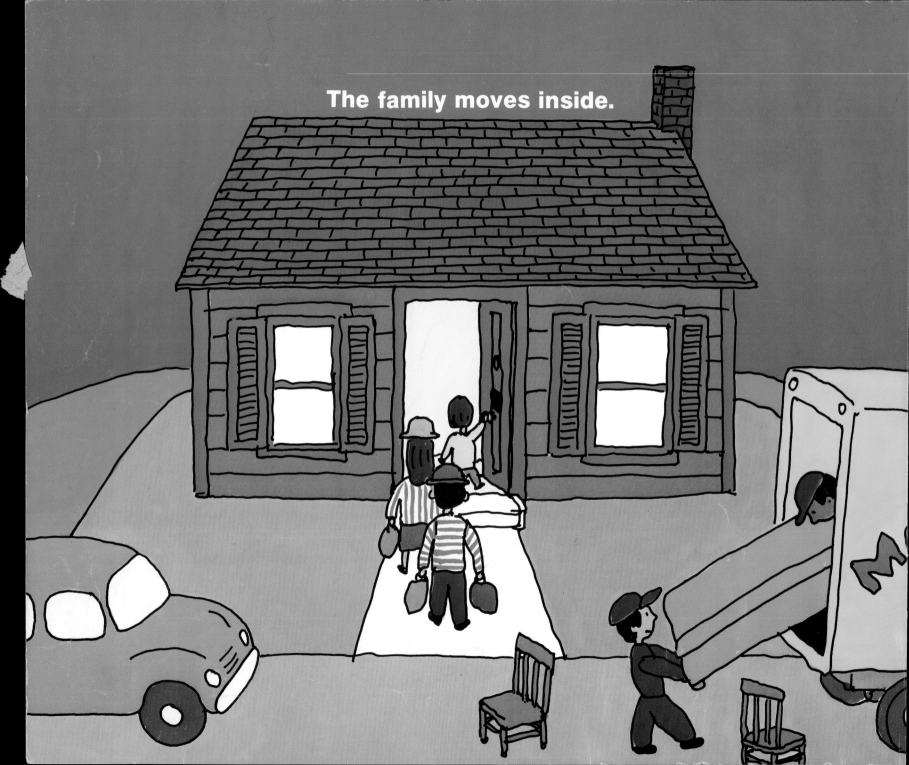

The family moves inside.